THE SOLUTION TO EVOLUTION

A questioning, scientific, study disputing the Darwinian ideas for the origin of the species. These ideas are logically challenged for proof and credibility. Also presented is the new hypothesis: THE SOLUTION TO EVOLUTION. This is a simple and beautiful hypothesis with supporting evidence; a solution that, surprisingly, has been staring us all in the face.

Contents

1. DID THE PRIMORDIAL SOUP EVER EXIST? 1

   Did life descend from space? Everything grows bigger and bigger.

2. A CYCLE OF LIFE IS REVEALED 8

   The time of arrival on Earth explains size. Creatures never evolve to become smaller.

3. DID NEW SPECIES EVOLVE BY CHANCE FAVOURABLE VARIATIONS? 14

   Do we have any proof? The Peppered Moth not accepted.

4. THE SOLUTION TO EVOLUTION 22

   New species appear suddenly. Leaps of development.

5. DID MAMMALS RETURN TO THE SEA? 31

   Immature mammals. Why gills did not re-evolve.

6. WHAT CAN METAMORPHOSIS EXPLAIN? 38

   The chicken and the egg. The development of the eye. Marsupials. Simple to complex forms. The diversity of life. The Cambrian explosion.

7. MAN ALWAYS WALKED UPRIGHT 51

   Man was always intelligent. Could man have been more intelligent?

8. THE PREVIOUS FORM OF PRIMATES 56

   Amphibians take to the trees. The Tarsier.

9. WE DIDN'T EVOLVE FROM THE APES 62

Why man is so different. Neoteny.

10. THE SKELETON..68

The insect stage. Do we have one common ancestor?

11. THE FINAL METAMORPHOSIS 72

From the amphibians. The dinosaurs.

12. STRETCHING THE IMAGINATION................................77

Long beaked birds and long trumpeted flowers. Is mimicry really mimicry?

13. LIFE IS NO ACCIDENT..82

We come complete with instructions. A speck of dust capable of asking questions.

1. DID THE PRIMORDIAL SOUP EVER EXIST?

Did life descend from space? Everything grows bigger and bigger.

The Origin of Species by Charles Darwin was published in 1859. Although its publication ruffled many feathers it gradually became accepted as a scientific truth. Today, however, there is far less certainty about this theory. Yet it continues to persist. The main reason for this is that no one has come up with a better idea. In order to topple this theory you need to have an exceptional idea with plenty of evidence to support it. It is now 2014 and surely time to look again, scientifically and logically, at Darwin's idea of evolution and natural selection and to examine any claims given as supporting evidence. We have much more knowledge available to us today to evaluate possible processes of evolution, that it is not unreasonable to suggest, that the time is now ripe for a new hypothesis. We need a new hypothesis, not only to open up fresh ideas on biological evolution, but also to change the present well-established theory that is now preventing progress. So, let me take you on a journey of logical reasoning to discover the truth and ultimately the solution to evolution.

Darwin believed, and so did many before him, that all the species developed from simple life forms. I don't think we can have any argument with this line of reasoning. It is, after all, quite logical because no animal as large as an elephant or, come to that, even as large as a mouse could suddenly spring into existence. Life just had to start with simple organisms that gradually developed and grew larger over time. There is no other way. Scientists all seem to agree on this one.

According to Darwin, everything began in a 'warm little pond.' Other sources now refer to this birth place for life as the 'primordial soup.' The story goes that this warm soup was rich with dissolved minerals, complex chemicals and electric charges emitted from the atmosphere. Life as simple cells is believed to have originated here as a chance, once only occurrence. Can this really be true? Well, we are told that the odds for chemical processes coming together in this way and producing life in Earth's primordial soup is such a remote chance that, if it happened at all, it certainly would never happen again; not here or anywhere else in the universe. If the odds really are that great then it's highly probable it didn't happen on Earth either. If life had originated on Earth it would have needed protection from the natural forces at that time. For example, protection from drying out, intense heat, radiation etc., so a primordial soup would have been essential. However, we cannot be sure if that primordial soup ever existed and therefore, whether life started here on Earth or came from elsewhere. Can we ever know? Well, it is surprising how sometimes we find answers, to what seems completely out of our reach, when tackling a different question. So, just for the moment we will put this one on hold and come back to it later.

What other scientific hypotheses for the origin of life have been put forward? There is the hydrothermal vent theory. Some scientists believe that these vents may have been the source of life on Earth. These vents deep on the ocean floor pump out chemicals and heat into the water. Apparently bacteria have been found around these vents that can cope with the high temperatures and thrive on the chemicals. However, before we get too excited, it doesn't prove that life was created here, only that it is happily thriving here. In fact, all we have done is to swap our warm pond or primordial soup for a hot vent. This means we are left with the same problem of a highly unlikely event. The only other theory presented to us, is that life descended to Earth from out of space. Where in space would it form? Where would it be coming from? Sir Fred Hoyle and Professor Wickramasinghe, believed that life descended to Earth in the form of bacteria. They were convinced that bacteria could be found in interstellar clouds. At first, it seems that this theory is just shifting the whole problem out of reach but let's take a closer look. If microscopic life had drifted down to the surface, at the dawn of our planet, it would have perished in the extreme heat and volcanic activity of that time. However, as conditions improved, later arrivals would have been able to survive and evolve. We can talk about 'later arrivals' because we don't have the problem of the 'once only occurrence' that we have with the chance event in the primordial soup. Can we prove or disprove this theory? Yes, because if this theory is correct then it should be possible for us to find microscopic life still drifting down upon the Earth today. If we can find evidence of this, we can prove this theory. In fact, microscopic life forms are indeed around us today but we are led to believe that they are the descendants from that 'once only occurrence' in the primordial soup. Now why would they not have changed and evolved in any way from that day to this? How can it be true? All microscopic life had to evolve, and to increase in size over time, in order to produce the creatures we see around us today. If they had not evolved and increased in size with every generation we would still be the size of microbes. As we said earlier, scientists all seem to agree that life started as simple cells and gradually developed and grew larger over time. If we believe that we have evolved, from microscopic life, then how can we also believe that microscopic life hasn't changed in any way since its creation in that bowl of soup billions of years ago?

In light of the above development we now need to go back briefly to that primordial soup and our unanswered question. Scientists tell us that conditions on early Earth eventually changed, particularly with an increasing presence of oxygen, and that this would have been toxic to the formation of new life. If this is true, then at that point in time, the life-giving primordial soup would have been removed from the menu. This would have meant that no new life could arise and take hold on Earth even if it had been possible for such an event to happen more than once. So, we would be left with only that successful, first formation of life, that started up before the advent of oxygen in the atmosphere, to develop and adapt to the changing environment. In fact, had this initial life actually occurred, it would indeed have been the once only occurrence. However, this story is not verified by the evidence we see today because microscopic life is still all around us and, as we shall see later, it is still capable of setting up new life on Earth. So, where is it coming from if not from out of space? It would certainly not be from a primordial soup because this is no longer available and certainly not descendants from that first successful population because, had they existed, they would have evolved and grown larger long ago. We now have the answer to that impossible question - did the first early forms of life start on Earth? No, life came from elsewhere. As we continue, it will become more obvious that the primordial soup and its birth of microscopic life never existed. Perhaps now is the time to think briefly about our amazing Mars landings by unmanned spacecrafts and our desire to find evidence of early life. If we do find evidence, that early Mars once supported microscopic life, what will be the general conclusion? Will it be proclaimed as proof that life arose more than once or proof that microscopic life comes from space and drifts down upon most planets?

Now that we are not floundering in our 'little warm pond' we can examine the outcome of this new freedom. We have microscopic life around us that has not yet evolved so we can be sure that new life is still descending from space. Wherever or whatever its source in space, it is still churning it out as it must have been doing for billions of years. We must assume that, where conditions allow, this microscopic life will take hold and develop. We can even play with the ideas that inevitably come to mind. For example, although it may sound like science fiction, if this planet remained in existence for long enough and conditions remained hospitable for their development, this new life might yet again produce those creatures that have long been extinct. This line of thought will be pursued in another chapter where we will explore this possibility of life starting up on Earth more than once and generating recurring species. The acceptance of the idea that microscopic life has a cosmic origin and is still arriving and evolving is extremely important because it supports the existence of life elsewhere in the universe. Obviously the same microscopic life would be drifting amongst most planetary bodies and although most will inevitably perish, in a vast and hostile universe, some will find conditions to support their existence. Where suitable conditions continue for long enough, life will be able to evolve as on Earth. If life is built from the same DNA, as this system of dispersal suggests, then the strange fictional life forms we imagine will not exist; our alien contacts will be human.

2. A CYCLE OF LIFE IS REVEALED

The time of arrival on Earth explains size.
Creatures never evolve to become smaller.

We have removed the Earth bound idea of the life-giving primordial soup, because it does not stand up to scrutiny, and have come to the conclusion that life originated somewhere in space. We arrived at this decision because 'life from space' has evidence to support it whereas there is a complete lack of evidence for life beginning in a primordial soup. If we continue to pursue this line of thought we reveal a 'cycle of life.' -- Microscopic life drifting down to Earth, followed by developing, evolving, increasing in size, and eventual extinction, while a new drift of microscopic life begins the cycle all over again. Following on still further with this idea, it amazingly reveals that therefore, there must be a relationship between the sizes of creatures around us today and the time of their arrival on Earth as microscopic life forms. Therefore, the microscopic life forms that survived to become the very early precursors of the blue whales would have arrived on Earth long before the microscopic life forms destined to be our modern elephants. The microscopic life that led to humans would not have arrived on Earth until well after the elephants and the tiny shrews not until well after humans. We don't have to guess that each new drift of life eventually evolves to be the same or very similar, to the creatures that came before them, because we have the evidence we require here on Earth. One look at our record of the animals that have ever existed on our planet and there is no doubt. If we consider some of the cat family, where similarities are obvious, we have in order of size the extinct machairodus Kabir, smilodon (sabre-toothed cat,) homotherium and today's tiger, lion, jaguar, leopard, puma, cheetah and many others right down to the domestic cat. They are not, as we are told, bigger or smaller because they chose to be, or because they evolved to become smaller, or because great size is an advantage but because their lives as recurring species began on this planet at different times. Those that came first would obviously be bigger than those that came later. We are looking at growth over time. Sir Fred Hoyle and Professor Wickramasinghe, believed that without doubt this planet has mothered many recurring species.

This cycle of life also supports Cope's Law. Cope's Law that basically asserts that everything continues to get bigger and bigger is shown to be correct. Edward Drinker Cope was unjustly mocked for his observations. The often quoted lines that dismissed his idea as nonsense are: 'There are far more small varieties than large ones today' and 'Some creatures, that we know from fossils were huge, have evolved to become smaller.' Using our method of reasoning we can see that there are many smaller creatures because they are all the generations that are developing and gradually increasing in size from later arrivals of microscopic life to Earth. Larger animals would be less in number, not only because they produce less offspring, but because they also require more food to survive and have less opportunity for sheltering from the elements and for various other reasons may become extinct. However, what should have been dismissed as nonsense is: 'some creatures have evolved to become smaller.' This is not logical. Nothing goes backwards. We have already agreed that life evolves to become bigger and bigger. We don't find that comparing ourselves to our grandparents or great-grandparents that we are gradually becoming smaller with each generation. What we do see is quite the reverse. Our acceptance of the Cycle of Life, with new life descending to Earth and starting the process all over again, clearly illustrates why we had some large animals in the past that are now represented as smaller. The large animals that existed in the past were once small but gradually grew larger and larger with time and eventually became extinct. However, during their lifetime more batches of new microscopic life had been arriving on Earth. These developed and gradually grew to look the same or similar to those that had become extinct. Some even later batches of microscopic life have been gradually developing and have reached the various sizes we see today; ultimately they too will reach a very large size before becoming extinct. They are the same type of animal but are not direct descendants as they developed from later arrivals of microscopic life to Earth. As they gradually increase in size, fresh batches of microscopic life will continue to descend to Earth to start the cycle all over again. For how long will the cycle continue? Will the flow of new life eventually ebb to a close, leaving this world and the universe totally devoid of life, or is there, as they say, more to life than this world dreams of?

We are told that present day horseshoe crabs are the descendants of their kind from 250 million years ago and that somehow their ancestors had managed to survive the great extinction of the cretaceous period. This cannot be true. Even if they had survived the cretaceous extinction they would still have grown huge and died out millions of years ago. The horseshoe crabs we see today are the descendants of much later arrivals of microscopic life to Earth even though they look like living fossils that haven't changed since the cretaceous period.

If we are right in our investigations so far, then it has revealed a most uncomfortable fact. The continual increase in size over a long period of time means that eventually creatures of great size must become extinct and are replaced by recurring species. Therefore, unless man can find a way to limit his growth he will follow the same fate. He will become extinct and all his endeavours will be lost. Is this why the heavens appear to be so quiet when we try to pick up signals from civilisations millions of years older than our own? It is certainly a sombre thought.

An article in National Geographic (April 2005) reported the discovery of small human skeletons, 'Homo Floresiensis' just over three feet tall, (nicknamed The Hobbits) that had evolved to become smaller. But, as we have said previously, nothing evolves to become smaller. If they really are human skeletons we have two possibilities. They could be extremely old bones from a time when humans were smaller because we didn't start off at around six feet tall. But, as the report goes on to say that Homo Floresiensis lived just 18,000 years ago when modern humans were on the move around the globe it rules this one out. The other possibility (and the evidence that we need) is that these are a race of humans whose origins didn't start until long after modern man; from another line of development. (i.e. From a later drift of microscopic life.) This immediately raises another question. Could there possibly have been humans that walked this Earth many millions of years before us; humans that were in existence long before the arrival of the microscopic life that led to modern man? It is certainly an intriguing question. These humans, had they existed, might have walked with the dinosaurs but we have not found their skeletons or any other evidence. If there were humans that preceded us, we can only conclude that they must have perished early in their development, like the much later Homo Floresiensis.

Sir Fred Hoyle and Professor Wickramasinghe, came to the conclusion that life came to Earth from out of space and in this way can re-establish itself. In their book, 'Evolution from Space', they state: 'In our view life on the Earth probably had many beginnings. With living cells from cometary sources constantly showered over the Earth's surface, many would manage to survive temporarily, later to become extinct, and then, perhaps to become re-established again. Thus we are not committed to a single line of development.'

3. DID NEW SPECIES EVOLVE BY CHANCE FAVOURABLE VARIATIONS?

Do we have any proof? The Peppered Moth not accepted.

Darwin believed that man had developed from primordial cells, through fish, amphibians and mammals and an ape-like creature. He thought that 'natural selection' was the main means of modification and that it acted slowly by accumulating slight, favourable variations. However, does this really work? Obviously sickly creatures will be eliminated but if a creature should acquire a small random change, for example, longer ears or neck or even a slightly different shaped beak, how can we possibly say whether this new acquisition will enhance or reduce its success? If we apply our usual logical reasoning then we would have to say that surely any such changes, should they occur, would have both advantages and disadvantages. What is obvious in the natural world is that creatures make the best possible use of what they have. They find suitable niches for themselves and flourish. Thus they would always turn any change to their advantage.

Creatures, whether large or small, appear complete in every way. All of them are living systems where every part connects and functions with every other part to make up the whole body. In no way is it apparent that any part of their system requires any 'favourable variation' to increase performance. Come to that, it is difficult to see how any such intrusion could connect into such a system without causing havoc.

Any slight favourable variation, should it occur in an individual, would need to be passed on to the next generation if natural selection is to work. Within a large breeding population these would be watered down with subsequent generations and probably disappear. In a small isolated population, where inbreeding is inevitable, it would be more likely to appear in the young. Nevertheless, so would the possibility of fatal inherited disorders that could wipe out a small population. Darwin's finches on their small island would only have needed a single bird to suffer from the genetic error of a twisted beak, for the inbreeding in such a small community, to pass on the error to other birds. Fortunately, these birds made a possible disadvantage into an advantage by finding food they could access with their type of beak.

However, it would be highly illogical to claim that natural selection is responsible for the evolution of species. Our reasoning tells us that the process is far too weak, too unreliable and depends far too much on 'chance' events. Also, the sheer number of favourable variations required to change one animal into another make it impossible.

The modern day interpretation of the above, that we read so often today, is that all new species arose by 'random chance mutations' over millions of years and that primates, including man, have evolved in this way from tree-shrew ancestors. Does this really make sense? It appears that many ideas have been supported by so-called chance events.   The cells coming together in the primordial soup were, we are told, a once only chance occurrence. The changes made to create new species were made by chance mutations. That intelligent life managed to evolve was pure chance. We need to look closely at ideas and explanations that rely solely on chance. Anything explained by 'chance events' is probably an easy way to fill an unknown void and quite probably never happened. We have already dismissed the idea of the primordial soup and the once only chance occurrence of life so we will now tackle the other two. That intelligent life was down to pure chance is a fanciful story because true facts are not available. You only have to consider the complexity of the human brain to realise that its development could never be attributed to pure chance events. Still, the fact that the brain did 'develop' at some stage, (and that is quite different to changing by chance mutations) is quite clear and we will be coming back to this later. So, what about the big one? 'New species evolved by chance mutations.' This is the one that makes everyone squawk. Obviously this is a very unsatisfactory hypothesis but if it isn't correct then we must find another explanation. There is always the one you haven't thought of -- the one where you find yourself saying, now why didn't I think of that?

If we do accept this theory of changes by slight mutations over millions of years, it means we have to believe that the growth of small body parts, or small changes to body parts, gradually accumulated and changed one creature into another. If a shrew gradually changed into a human, as many would have us believe, think of the number of chance mutations that would be required. In fact, you begin to wonder if you can call them 'chance events' when you need so many of them to make the transformation and particularly when they all have to happen in the right place as well. When you weigh it all up, I think we would still be looking more like a shrew today, don't you?

Do we have any proof that new species came about by chance mutations, favourable variations and natural selection? The answer is no, we do not. With the complete lack of transitional forms in the fossil record it cannot be proved. Although there have been some desperate attempts to seek out creatures that show such missing links, they are as unsatisfactory as the idea itself. Let us consider the horses that are claimed to show this progression. We read that there is some dispute about the fossils not appearing in the correct order in the rock strata, some it seems were living at the same time as each other and there is also doubt about whether the first horse, Eohippus, was really a horse. We cannot judge the truth of these statements but it is quite obvious that our main objection must be that these horses do not show a progression in size. Some horses shown in the line of development are smaller than their ancestors. We have already established that creatures do not evolve to become smaller so we cannot accept this as a missing link.

Another example is the 'Peppered Moth' that is quoted so often as an example of evolution in action. It is stated that this moth, originally mottled white and black, was camouflaged on the lichen covered tree trunks on which it rested. However, in industrial regions the lichen was apparently killed by the pollution and the tree trunks were covered in soot. This gave better camouflage to a mutant black form. Therefore, the story goes that the white ones, that were no longer camouflaged, were eaten by birds and the black camouflaged ones were left to flourish. There is something here that doesn't sound quite logical. For one thing there are still many healthy populations of white ones. So, why haven't we ended up with all black ones when these were left to breed undisturbed while the birds were devouring all the white ones? Birds have excellent eyesight so it's unlikely that they would be fooled as easily as we are. Also they would only have to come across a black one that moved and they would immediately be on the lookout for more. If you were desperate to prove a point though, it would be easy to fall into a trap here without looking for any other explanation. Did you know, for example, that butterfly breeders can radically alter the appearance of butterflies by keeping larvae and pupae in a refrigerator? Apparently a difference in temperature can make them look like different species. Therefore, butterflies emerging in the spring (or the refrigerator) would look much whiter than butterflies emerging in the warmer summer. (These would be blacker.) Now the point here is that industrial areas would also be warmer than elsewhere and therefore, create blacker forms. If you remember, we were told that this only happened in industrial areas and was attributed to the soot being the perfect camouflage for the mutant black form. If global warming is happening then maybe they will all turn black. Some information on the Tiger Swallowtail butterfly is of particular interest here. The female of the Tiger Swallowtail butterfly has two forms, one has yellow colouring like the male and one is black. As the black ones are mainly in the south it looks like the result of a warm temperature again. Then there is the Map butterfly that has seasonal variability. It has two generations, hatched in different seasons of the year, where the summer (warmer again) female is dark and the spring female light in colour. Questions immediately jump out at us here; in particular has anyone checked if the Peppered moths with black colouring are all female? Therefore, it is possible that the changes have nothing to do with camouflage and sooty tree trunks, or indeed, evolution in action. Without further investigation we cannot accept these accounts of the Peppered moth as proof of evolution in action.

So, where does all this leave us? Well, we have to say, it leaves us without any evidence to support the idea that changes to new species were made by chance mutations, favourable variations, natural selection or whatever else you like to call it, over millions of years. We have obviously strayed down the wrong road on this one. This is even more obvious when we consider our continuing 'cycle of life.' This new life arriving and evolving will eventually become the same, or very similar, to the species we see around us today and the evidence for this can be found by looking at the complete history of animal life on Earth. Therefore, the hypothesis claiming that new species are formed by chance mutations just cannot be correct because it would be impossible for chance mutations to produce the same forms again. So, coupled with the fact that there is no evidence in the fossil record of transitional forms means we have no choice but to topple this hypothesis.

To topple this hypothesis we need, as mentioned before, to have an alternative explanation for the origin of the species; an explanation that is all encompassing and beautiful and leaves no room for doubt. Any attempt to topple the present theory will cause a major upheaval. It will certainly ruffle everyone's feathers again but at least we will be free to have a look at the signposts afresh and to take a new path. We are, after all, used to these reshuffles and recognise them as a necessary procedure for the progress of science. We once believed that the world was flat and once thought that the sun revolved around the Earth and that the Earth was the centre of the universe. Now, unfortunately, it seems there is a strong belief that life on Earth was a once only occurrence and exists nowhere else in the universe.   Our great-great-grandchildren will probably be amazed that we seriously believed humans were created as a species by chance mutations over millions of years from a shrew-like creature. However, let's not be too smug about it, for the truth is looking to be even more unbelievable.

4. THE SOLUTION TO EVOLUTION

New species appear suddenly. Leaps of development.

The geological record shows quite clearly that new species appear suddenly without any transitional forms. This absence is extremely important to our investigation. They can't suddenly spring into existence so how can we explain what appears to be a sudden change?

As we are now talking about sudden changes into a new species (not small changes over millions of years) it ought to be possible to observe how some of these small creatures are managing to evolve into new species. It stands to reason that such creatures would be extremely small and undeveloped because a sudden change for large animals would be impossible. So, when we look around do we observe any very small creatures on this planet making sudden changes into something new? Yes, we do. How are they doing it? The answer is by metamorphosis. So, here we have it--- 'The Solution to Evolution.'

What is amazing about this hypothesis is that we can still see it in action today. There are many examples but the two that are best known are the tadpole changing into a frog and the caterpillar into a butterfly. If butterflies gave birth to young butterflies, it would be considered utter nonsense to even suggest that a caterpillar could change completely into this beautiful winged creature, particularly over such a short period of time; but it is true! There is no gradual change or slow evolving of wings over vast periods of time. The butterfly is complete from metamorphosis. We are able to see that – *'A new life form or species, straight from its metamorphosis, establishes itself quickly as a functioning, workable body. It uses whatever it has been endowed with, to the best of its ability, and in whatever way comes most easily and comfortably to it.'* In this way a caterpillar changed to a butterfly functions quickly and easily within the scope of its new body. When a tadpole becomes a frog its legs have not evolved by slight mutations and gradual change over millions of years. The change from gills to lungs did not have to wait for chance mutations. It is a sudden event by metamorphosis. Metamorphosis enables leaps of development. Therefore, why do we imagine that the legs, wings or lungs of other creatures have taken millions of years to evolve from slight mutations? It can only be that these conclusions have come about when struggling to explain evolutionary changes happening in large animals, which of course, is impossible. These changes are all activated during the metamorphosis of very small, undeveloped creatures as they become the new species. Once this has taken place - you are what you are - except for a gradual increase in size over time. When we start to think about metamorphosis, as a means of species change, we can see that this is the only way that significant and complex changes can be made. In this way all related and connected body parts are put in place at the same time to create a fully functioning life form and this is what we see in all creatures; there is nothing disconnected about them. Their bodies are completely functional as a whole system, with each part only functioning by its connection to every other part. Therefore, slight changes made by mutations over millions of years could never hope to achieve this. To change or modify a complete system by mutations is impossible. The only way it can be done is through metamorphosis and only then by small, undeveloped creatures still capable of metamorphosis into new forms or species. The new species of animals then continue to increase in size over time.

We are beginning to see now, why it is that new microscopic life descending down to Earth ends up producing the same or very similar creatures time and time again. It must happen because they come equipped with the required DNA to make the whole series of metamorphic changes. In fact, it must be a continuous line of development, with only sections of DNA coming into play at any one stage. Each stage would be followed by periods of time for growth and development. Thus all the instructions for each body plan and development, including whole stages to be made by metamorphic leaps come complete. This would possibly also include the ability for completed parts to trigger the next on line. Each metamorphic change would continue from where the last one left off. In this way everything is in readiness for the changes and all the necessary parts for development will link up and function. This could never happen by chance mutations over millions of years. In fact, it makes us realise how impossible it would be for chance mutations to fulfil such a programme. Even for metamorphosis it is a miracle of life.

The lack of in-between species or so-called missing links can be explained by metamorphic change. Once a creature has changed through its final metamorphosis its previous form will cease to exist and the completely new species will begin. You might well say, 'Hey! Hang on a minute the caterpillar and the tadpole haven't ceased to exist.' That is because the caterpillar and the tadpole still have to make a complete metamorphic change to a new species and not just a change into a form that is solely for reproduction. How can we tell? We can tell by the fact that their present metamorphosis produces creatures that still have to give birth to their previous form and not replicas of themselves. Even with metamorphosis the changes cannot be made all in one go. Therefore, it can be seen that all the undeveloped small creatures from simple life to amphibians progress in stages through metamorphosis. Each time, a perfectly functioning creature is produced. Amphibians are the last of the small, undeveloped creatures to be capable of metamorphosis. Creatures that are the result of a final metamorphic change will give birth to exact copies of themselves and not the young of their previous form. After the final metamorphic change, - you are what you are. Therefore, once changed to the final species, its prevlous form will cease to exist (because this form is no longer being produced) and the life of the completely new species will begin and will give birth to copies of itself. The only change from then on will be to gradually increase in size. If you think about it, this is exactly what we see in the geological record.

Metamorphosis occurs to give birth to a new species but it also produces a viable form for reproduction when it cannot carry its development right through to the new species. For example, the caterpillar cannot lay eggs to produce more caterpillars. Therefore, it goes through metamorphic change to a form for reproduction (the butterfly). The whole purpose of the butterfly stage is a temporary form to reproduce the caterpillar. The wings aid this procedure by widening the area for seeking a mate and finding the necessary plant for its eggs and caterpillars. Eventually the caterpillar will metamorphose into its new form and not into the butterfly. When a new species is the outcome of a metamorphic change, it will produce exact copies of itself and never metamorphose again. Therefore, we must accept that metamorphosis does not occur in mammals of any size because they are the outcome of a very last metamorphic change. This, of course, dismisses any ideas that suggest one mammal can change into another mammal by small random changes or by any other means. So, we have to say, once a shrew always a shrew or once an ape always an ape.

There is another good example where metamorphosis produces a special stage for breeding when it cannot carry its development right through to the new species. In the sea, we find free-swimming life that is forced to metamorphose into a static plant-like existence. This is a sexually mature state for reproduction and does not produce copies of itself but of its free-swimming form. Eventually when size, nutrient intake and conditions etc are right, it will continue development from its free swimming form, becoming sexually mature and breeding in its own right. The static plant-like state will be bypassed. Metamorphosis is always triggered in the previous form of existence, (i.e. tadpole, caterpillar, etc., that often have unique structures and organs not found in the reproductive form) and is not triggered in the frog or butterfly, which is a form for reproduction only. This is particularly clear in some insects such as the Mayfly because it has no mouthpart to feed. The first stages of a metamorphic change may be much the same as before, but parts not required are bypassed as it progresses on with new development to produce the new species.

Some parts from metamorphosis are necessarily advanced in readiness for the next stage of development. These are not only comfortably accepted and adjusted to, but are positively utilised by its owner. For example, the frog as we know it today is faced with hind legs that are far too long. (Eventually metamorphosis, triggered in the tadpole, will continue on further with developing and making changes to what would have been the frog stage. This further development will be explained in another chapter.) So the frog uses those long legs in the way that comes most easily and comfortably to it and this is to leap rather than walk or run awkwardly on all fours. Therefore, the frog can make full use of those long back legs for leaping away from its predators. It can be seen that development through each stage of metamorphosis progresses in such a way that it is not a hindrance or useless to its owner but of positive benefit. At present we are told that the frog's long hind legs have slowly evolved over millions of years to enable it to leap away from its predators and that only those with the longest legs survived. We can see now that this is not true. Indeed, if it had managed to survive for millions of years while its legs were gradually growing longer then it didn't really need them to be longer anyway.

Before we can continue with our investigation we need to consider what triggers metamorphosis. Development is one point to consider since all creatures mature over time. Another could be access to certain nutrients. For example, we know that tadpoles cannot turn into frogs without sufficient thyroxine. Climatic conditions and particularly temperature must also be important because we know how this affects life in general. We see the bursting of new life once the warmth of spring arrives and the necessity of warmth for the incubation of eggs, the rearing of young and the growth of vegetation upon which we all depend. In alligators the warmth of the eggs even determines the sex of the young alligators. The level of light could also be involved and so too could moisture levels. Whether it is one or many of these conditions might become clearer as we proceed. What we can observe from present creatures going through metamorphosis is that the instructions are triggered throughout a species at more or less the same time. This is essential because it provides a significant gene pool for breeding and carrying on the species. If we had to rely on small random changes, happening over millions of years, it is difficult to see how a significant number of creatures would be involved to create a sizeable gene pool capable of passing on their newly acquired characteristics.

Another point here is that the sudden triggering of metamorphic change could produce plague proportions of new species on an unprepared environment. Larger creatures could then become extinct due to temporary imbalance of nature within the area. This might take the form of depletion of food, competition for habitat, and possibly intimidation affecting the rearing of young. Thus, the fossil record would show not only the sudden appearance of a new species and the lack of any obvious previous form but also the extinction of other creatures as well. This is exactly what we do see in the geological record.

It is also interesting to contemplate the possible effects of global warming. What metamorphic changes could be triggered by the rise in temperature? What new species could suddenly plague the Earth and create utter devastation for humans? Would they present a danger to man or to the crops and animals upon which he is dependent? Supporting evidence for a warm climate triggering metamorphic change is found between the lines of a report in New Scientist (22 June 2002). The article from the proceedings of the National Academy stated that carbon dioxide could speed up evolution. Apparently, comparing the CO_2 levels with the fossil data saw a close match between high CO_2 levels and more new species. This is interesting because CO_2 is responsible for the greenhouse effect. Therefore, we can see that the subsequent warming up of the planet, rather than the carbon dioxide itself, would have been the trigger for metamorphosis and result in the rise of new species.

5. DID MAMMALS RETURN TO THE SEA?

Immature mammals. Why gills did not re-evolve.

We read over and over again about mammals returning to the sea and losing their legs as they changed to fit that environment. Does this really sound feasible? Would any creature used to the freedom and mobility of legs choose to abandon the land for the sea? To have fully developed legs instead of flippers and a suitable tail would be an exhausting struggle in water for even a few hours, let alone for years of slow change. There is absolutely no evidence to support this theory that limbs, once obtained, can be reduced to mere vestiges if not used. A mammal, as we have said before, is the result of a last amphibian metamorphosis and once completed (at whatever stage) their bodies cannot be changed by any further additions or, indeed, deletions. Mammals that spend a great deal of their time in water to keep cool or to catch fish do not lose their legs. Hippopotamuses still have legs and if we consider the length of time it has taken them, as a species, to reach this huge size, we realise that animals do not lose their legs even if they spend most of their time in water. This is that old idea creeping in by the back door again. i.e. that behaviour can reduce limbs or body size or lengthen necks or beaks. It just isn't logical. Whalebone whales have no teeth, ear pinna or neck and a completely different tail to a land creature but they have lungs. If they really had returned to the sea it would have been more advantageous to lose their lungs and retain their teeth. After all, what do you eat between losing your teeth and developing baleen to sift krill and think what a disadvantage lungs are in the sea. It is interesting to note here that under the skin of whales' fins there are front leg bones and digits. Also, we can still detect what were once the tiny beginnings of back leg bones deep within their bodies and teeth can be detected in the whale embryos but afterwards disappear. Nevertheless, this is not evidence of a terrestrial past and that they once had legs and teeth and that these disappeared when they took up life in the sea. How do we explain it then? It is explained by metamorphosis. When whales were very tiny creatures, going through metamorphosis from their previous form (the amphibians), they became trapped in the middle of their metamorphic change due to severe climatic conditions. Although they were unable to continue their development, they were able to reach sexual maturity and breed at the stage they had reached. They continued to increase in size to the present day. As the whale's size increased, the bones which were the beginning of back legs remained small in relation to body size because the metamorphic process for legs had ceased. These bones are definitely not evidence that they returned to the sea. In fact, the reverse is true. It surely makes much more sense that long ago, when they were very small creatures, they were changing to come out of water onto land. They had developed lungs and were beginning to grow legs, much as tadpoles do today, but then due to cold climatic conditions their development ceased.

We know that extreme changes in climate can arrest metamorphosis. Intense cold would have stopped the metamorphic development of creatures at whatever stage they had reached. Certain requirements and climatic conditions are not only necessary to trigger the stage of metamorphosis but also to complete metamorphosis. Therefore, problems would have arisen if these were not available or the climate changed before metamorphosis was complete. Our planet has suffered many violent extremes of weather so it is not unreasonable to suppose that some creatures were caught in the process of change and locked to a life in the sea. So, what can this tell us about whales, seals, walruses, manatees, dugongs and other mammals in the sea? What would they have been? After all, it is believed that sirenians share a common ancestry with elephants, and cetaceans with hippos. (We can imagine what might have been produced had metamorphosis continued.) All of these creatures are, in our eyes, distinct species but in reality they are the incomplete forms of some of our familiar land creatures.

Richard Dawkins wonders in his book, The Greatest Show on Earth, why whales, dolphins, dugongs and manatees have not re-evolved functional gills. He states: 'The fact that, like all mammals, they have, in the pharyngeal arches, the embryonic scaffolding to grow gills suggests that it should not be too difficult to do so. I don't know why they haven't but I'm pretty sure there's a good reason, and somebody either knows it or knows how to research it.' If you believe that these were once land mammals that returned to the sea, you would indeed wonder why they have not re-evolved gills. However, once you realise that these creatures were long ago evolving to come out of the sea, it becomes perfectly clear. Their lungs were developed during metamorphosis in order to come out of the sea and their gills were reabsorbed. The embryonic scaffolding is all that remains. This is a good example of how a little bit of wrong information not only raises those questions that can never be answered but can also send you off on a wild goose chase.

Further evidence is shown by the flippers of true seals, and walruses etc. These have internal hand and finger bones and also arm bones although greatly shortened. These bones are completely enveloped in the skin of the flipper. In embryo development an arm bud grows much like a flipper without any division between the fingers until finally, at a certain stage, the skin retracts from between the fingers. Even the human embryo starts off with hands like small paddles but then, as a certain stage of development is reached, the skin retracts from between the fingers. This is clear evidence that these mammals in the sea are immature mammals and that their ancestors were once in the process of developing to come out of the sea. They were developing by metamorphosis but never reached the stage where the skin retracts from between the fingers, or indeed, any other changes that should have happened at that time to enable them to reach their full potential as land mammals. In fact, we have the evidence for cessation of metamorphosis at different stages in sea mammals. Some have no sign of hind legs externally. The whale, the dugongs and manatees have beginnings of hind limb bones floating deep within their bodies. Others have weak, undeveloped hind leg and feet bones, encased in flippers, that are of little use on land and they have to drag themselves on shore with their front flippers. Some have hind leg bones and feet that show more development within the flipper and can support the body. Further evidence is the lack of external ears (with the exception of the sea lions and fur seals and that makes sense because they also have more development of leg bones and feet within the flippers). that shows that their metamorphosis stopped before this stage was reached. Need we add that lack of external ears does not show that their ears gradually grew smaller over millions of years and eventually disappeared?

Snakes are another example for creatures that were unable to complete metamorphosis. No trace has ever been found of front legs in snakes, only back legs. This would suggest that the back legs started to grow but then metamorphosis was arrested. Pythons have claws on their back legs but this is the only part that is visible externally. The snake, just like the whale, did not lose its legs; it never managed to complete development. If their metamorphosis had reached its full potential we would not have snakes today but a different creature with four legs.

One example that we can see today with metamorphic problems is an axolotl which is the larva of a salamander. It hardly ever manages to complete metamorphosis into the salamander and breeds in its larval stage. Again, cold conditions appear to be the problem although there might be other factors involved such as a lack of thyroxine. In its natural habitat it is living in cold water but when reared in a heated tank it manages to complete metamorphosis into the salamander. (It is possible that this might also be due to the low water level of the tank; a reaction to drying up.)

Metamorphosis, it would appear, is a risky business. A number of requirements have to be in place to trigger you off in the first place but you are at the mercy of the climate while you go through the process; and there is no turning back. It's possible that during extremes of cold weather on this planet that some creatures could not make metamorphosis. Also, it is clear that some were caught in the process of change. Unable to complete their development they were left to make the best possible use of their half-way stage. So, unable to go back to their old form and blocked from completing metamorphosis into their new form, these creatures would have become sexually mature and produced young with the same incomplete development.

We can only assume that where life was able to become established on other planets, throughout our galaxy, it would also have been at the mercy of the climatic conditions of its host planet. So, where climatic conditions were ideal, then life would have been able to complete metamorphosis without any problems and there would not be incomplete mammals in the sea. In fact, many of the creatures that we have on Earth would not exist on an ideal planet because metamorphosis would have reached completion and would not have ceased at various premature stages.

6. WHAT CAN METAMORPHOSIS EXPLAIN?

The chicken and the egg. The development of the eye. Marsupials. Simple to complex forms. The diversity of life. The Cambrian explosion.

Metamorphosis can explain how the chicken came before the egg. If new species were not born through metamorphosis you would have the problem of which came first, 'the chicken or the egg.' If the egg came first we end up with the problem of where it came from because we need the chicken to lay the egg. If the chicken came first, then where did it come from because a chicken comes out of an egg? This strange circle shows that we definitely have our facts wrong somewhere. Now, if we apply the theory of metamorphosis, as the solution to evolution, the problem does not arise. The first birds would emerge from the metamorphosis of their previous stage and although extremely tiny they would nevertheless be adult birds. The emergence from metamorphosis is always in the adult and sexually mature state for reproduction. They would already have wings and be able to fly. There would be no need for millions of years to slowly evolve wings for flight or with flapping around with part-wings or flaps of skin for gliding until flight was gradually achieved. There never was a stage of part-wings or pre-flight. The bee humming bird is only 5cm in length but has wings and can fly. The metamorphosis to birds was in some ways unusual and incomplete. This is shown by the unusual development of the hand and finger bones, the incomplete development of the fibula and ankle bones, no external ear, no teeth and they were still egg-layers and retained the cloaca. (Although they did have a four-chambered heart like a mammal.) We need to be able to explain this but for the moment we can only surmise that they must have been sufficiently developed beyond their previous stage to be able to produce copies of themselves (i.e. their eggs did not produce their previous form). Out of the egg came the first young of the new species. So, we have it in a nutshell or, if you like, an eggshell –the answer is simple; the chicken came before the egg.

If Darwin was right about the origin of species we wouldn't be seeing metamorphosis to frogs, butterflies etc., because there would be no need for metamorphosis when changes could come about by mutations and gradual change over thousands of years. So, does life make changes by mutations or does life make changes by metamorphosis? You can't have it both ways. The fact that we are able to observe metamorphic changes on our planet shows quite clearly which idea is right. We are most fortunate that we can still observe some creatures undergoing metamorphosis because without them the origin of species would have remained a puzzle for ever. Who could have imagined that life comes programmed for development in stages, implemented by its DNA, and that it can change so dramatically and rapidly in this way?

The development of the eye has always been a problem for the theory of evolution, even apparently for Darwin. The problem mostly raised is that it is difficult to imagine how the eye could develop gradually over millions of years because part of an eye or a slowly developing eye would be useless to its owner. So, can metamorphosis provide a better explanation? Yes, because even metamorphosis cannot do everything in one go which is why creatures go through a number of metamorphoses. The same gene is responsible for the eye whether it is the eye of a fruit fly, a frog or a primate so the information is there for continued development. With every change by metamorphosis the eye takes another jump in its development and therefore, is fully functional at every stage. In the very simple forms of life we have the light sensitive spots. This is the earliest development of the eye which must have been a great asset to its owner. During the metamorphosis of these creatures into their next stage (the insects) the eye is developed into the compound eye. As we can observe today, this provides efficient and adequate sight for them. During further metamorphosis into new forms the advanced eye is developed. Therefore, the eye is not a useless or part-functional organ at any point throughout its development but a complete miracle of vision to each stage of life emerging from metamorphosis. It is interesting to note that our vision is recorded on the back of our brain in strict vertical lines. Could this possibly be evidence of development from ancient structures such as the compound eye? Everyone can accept that the eye is a very complex structure so most find it difficult to fathom how it could possibly have been produced by random mutations over millions of years. There have been a few voices over time stating that a progression from primitive eye spots, to compound eyes, and then to advanced eyes, show how the eye must have evolved. Unfortunately, because no creature in the fossil record could show the eye in its in-between stages, it was not taken seriously. These voices were so close to the truth but without the idea that metamorphosis is responsible for species change and progression of the eye in jumps, they could not make the connection. While we are on the subject might we now ask which piece of the human body is not very complex? The ear is surely just as complex and what about the brain, the heart and circulatory system and so on and so on? In fact it is difficult, if not impossible, to see which part could easily be produced by mutations over millions of years. The same genes, but different anatomy, occur in other organs besides the eye; the heart, sense organs etc. This is because the DNA of a creature carries the information and instructions for both simple and complex forms, which means it is the same gene that is using further stages of information.

Marsupials present us with another puzzle. How do they fit in with the idea that metamorphosis is responsible for species change? It would appear that metamorphosis successfully developed them past the egg-laying stage (except for the monotremes) but failed to complete their development through to a placental mammal. Therefore, they would have been left without the ability to provide egg protection for foetal development and also without the ability to provide full-term protection in a womb with fully functioning placenta either. In marsupials the placenta is either absent or very rudimentary. The young are, therefore, born extremely prematurely in a foetal-like state and remarkably manage to crawl into the pouch. Here they attach themselves to a teat to continue their development. However, the added item of the pouch to enclose the teats (and protect the joey in its foetal-like state) does not appear to be explained by incomplete metamorphosis. We can understand a creature being left in the middle of a process but how do we explain this addition? Does life have a back-up system for every stage reached? It is interesting to note here that in the whale (an incomplete mammal) the mammary glands are enclosed in a sort of pocket or flap so the possibility of a pouch is in the system somewhere. (Not all marsupials have a deep pouch like the kangaroo. Some have simple flaps of skin over the nipples.) In fact, it shows up in some other creatures too. There are many South American frogs that have cleft-like dorsal pouches or fully enclosed Marsupium-type pouches (as in the Gastrotheca species). The male seahorse has a pouch. The Yapok is the only living marsupial in which both sexes have a pouch. (The

Tasmanian tiger is believed to have exhibited this as well.) Could the pouch have been a development for a water existence? Strangely, in the Platypus that does forage under water the pouch is absent. We could say, that's because it lays eggs, but the Echidna lays eggs and that has a pouch. So, pouched marsupials must be the result of metamorphosis ceasing just at the time when the egg-laying method was being discarded, to start development of the womb and live-bearing stage. Then there would no longer be any system in place for producing the shelled egg. (Apparently an eggshell does form transiently but is reabsorbed before live birth. Thus it produces the premature foetus without a shell.) However, the stage for the development of the womb had not been reached either or could not continue development due to the cessation of metamorphosis. Marsupials have many other differences to placental mammals. For example, the female reproductive organs have a different arrangement. There is a double reproductive tract with two uteri each having its own vagina. This suggests that continued development would have allowed time for these to fuse together as one; as apparently they do in placental mammals. Also they have a single opening called a cloaca. The male has the testes in front of the penis rather than behind as in placental mammals, with the only exception the rabbit (which is strange because it is not a marsupial). When metamorphosis ceased, the male's internal sexual apparatus would not have been able to continue its intended longer route and therefore the opening came at the rear of the scrotum for the penis. The continued longer route would have brought the penis In front of the scrotum as in placental mammals. Could it be that placental mammals and marsupials originated from a different line of arrivals on Earth rather than any changes during metamorphosis? This is not borne out by the fossil record where both have fossils in deposits of equal age. Can we tell by their size? No, because we can see that marsupials range from very small to quite large, even larger in the fossil record, which is exactly what we have in the placental mammals. It is now clear that marsupials were not a freak, one off development but over time have had a number of lines start up from new arrivals. Certainly we can dismiss the idea, prevailing at the moment, of placental mammals developing from marsupials because we have already worked out that mammals cannot change into other mammals. (Nevertheless, they were, in a sense, on the right track - see next paragraph.) We know that it is only small creatures capable of metamorphosis that can change into something else. Also it goes without saying that present marsupials, of any size, are not showing any signs of becoming placental mammals. Therefore, it looks like a cessation of metamorphosis at a critical time and climate is probably responsible when you consider where most marsupials were to be found.

Another question that is often raised is: Why are there marsupial mammals that look so similar to placental mammals? The answer, of course, is that they were going to be placental mammals (the same or very similar to our present placental mammals) but their development time during metamorphosis was cut short. Now as well, we can understand the strangeness of the rabbit. The rabbit is the only placental mammal with the penis behind the scrotum like the kangaroo and it also has the following similarities to the kangaroo: it has a cloaca, the female has duplex uteri only seen in rabbits and marsupials, the big toe is missing, and the forelimbs are relatively feeble. The head has a similar resemblance with the same long ears. It also uses its hind legs to hop rather than walk and will stand on its hind legs and box an opponent in the same way as a kangaroo. Both the rabbit and the kangaroo thump their hind feet to signal danger and both are crepuscular – that is most active near dawn and dusk. It would appear that the rabbit came very close to being a marsupial and not a placental but just a little bit more development made all the difference. However, once the rabbit has increased in size over a great many years, it will be recognised for what it is - not a rabbit but a kangaroo; a placental kangaroo. (Without a pouch and with its young more developed.)

The egg-laying monotremes (the Platypus and the Echidna) must have had an even earlier cessation of their metamorphosis. Their previous form would have been an egg-layer and this was retained because the stage of abandoning this for live-bearing had not been reached. It is understandable why the method of development by metamorphosis has suffered many setbacks with the climatic conditions and unrest of this planet. To name but a few we have had the ice-ages, movement of tectonic plates, rising sea levels, meteorites, volcanic eruptions, earthquakes, rising global temperatures, droughts etc. The platypus looks as if it's been made up from all the leftovers at the bottom of the pit. It's a mammal but it lays eggs, has mammary glands but lacks teats, has webbed feet, a snout shaped like a duck's bill, no teeth only horny plates, (it is born with teeth but these drop out at a very early age) no external ears, the rectum and urinogenital system open into a common cloaca, the shoulder girdle is very primitive, the males carry poisonous spurs on their ankles, and so on and so on. If we were all such a hotchpotch of body parts, then maybe we would be more than willing to accept that chance mutations have been responsible for the origin of species

Metamorphosis also explains the sudden appearance of new species. Creatures do not gradually change by mutations over millions of years; they appear suddenly. One day we see tadpoles swimming in a pond and next we see completely new creatures (frogs) hopping around on land. We are told that during the Cambrian period, 530 million years ago, many new creatures suddenly appeared and this is shown in the fossil record. We are also told that there are no apparent ancestors to these creatures in the earlier layers of rock. However, this explosion of new species can never be explained by looking for similar creatures in the previous strata because the creatures in the previous strata (Precambrian) will look completely different; as different as the tadpole and the frog. Thus in the Cambrian strata, (once metamorphosis had changed them all into their new forms and it would have happened rapidly) there would be no record left to explain their sudden appearance. The characteristics of these new species would have developed during metamorphosis and therefore, no recognisable earlier form of them will be found in the previous strata. Metamorphosis was responsible for generating these new creatures and it is also a biological process we can still observe in living populations today.

Metamorphosis can also explain the great diversity of life. Some metamorphic changes would not be triggered due to climatic conditions and this would leave creatures to continue unchanged. Sometimes it would complete right through to the new species. There would also be times when metamorphosis would cease during the process (and there is no going back) so that creatures would be left trapped at various stages of incompleteness. As any of these events could happen during any creature's metamorphosis we can see how this would create, over time, a multitude of seemingly new varieties that we would undoubtedly classify as different species.

Metamorphosis helps us to take the mystery out of what is referred to as convergent evolution. We are told that some creatures, that are not close relatives, independently evolved similar traits due to their similar environments.

This would be impossible if your only method of change was by random mutations over millions of years. Once we accept metamorphosis as the solution to evolution it becomes clear that creatures have not evolved similar traits as the result of their environment. After metamorphosis - you are what you are - and you make the best of your environment. It is quoted that the legless condition has evolved in snakes and also in several unrelated groups of lizards but we are now aware that this is due to early cessation of metamorphosis. It is also quoted that many of the Australian marsupials are almost identical to placental mammals living in other parts of the world. However, there is no mystery here because we now know that they would have been completely identical but for their early cessation of metamorphosis. We are told that the marsupials were only successful because the placental mammals had not reached Australia. However, it is now quite clear that this is not correct. Australia, like other parts of the world at that time, was in the process of producing its own placental mammals. The climate and harsh conditions in Australia caused metamorphosis to cease early and prevented completion of their development to placental mammals; this incomplete development resulted in the marsupials.

Metamorphosis shows us that life has purposeful direction through a sequence of development from simple to complex forms. The information for this must come within the DNA that is carried by the microscopic life forms as they continually descend upon our Earth. The continued cycle of life thus produces a sequence of simple to complex forms by metamorphosis and of course, without doubt, recurring species. The idea that life progresses from simple to complex forms was raised many times in the past but was always dismissed as incorrect. One continued criticism has been - 'If all living things are progressing, why are there still simple creatures left?' Through the system of metamorphic change we can see clearly that more advanced, complex forms are produced. The fact that simple creatures still exist is explained by the cycle of life where simple microscopic life is still descending upon the Earth and developing. These simple creatures will continue to develop and increase in size and, through metamorphosis, will evolve in complexity rapidly.

Can metamorphosis explain the development of the Bat? Its wings are not the same as those of birds. Even if we forget the feathers the anatomy is different. Both the bat and the bird have a skin stretching over arm and finger bones but the fingers are greatly elongated and support the wing in the bat. We know that during normal mammalian foetal development the fingers are enclosed by a membrane but at a certain stage of development this gradually retreats and leaves the fingers free. Therefore, this suggests there was a delay towards the end of completion of metamorphosis from its previous stage. The development into a tiny, furry, four-legged mammal did not complete at the right time and the fingers elongated and remained enclosed in the skin-membrane. Also some had extra growth continued at the end of the nose. We also have to consider the echolocation used by bats. It has been put forward that flying came before the echolocation but is this logically sound when we know that bats fly at night and frequent caves in total darkness. Surely, flying and echolocation would both be present at the moment of becoming a bat. As we have said before, once metamorphosis ceases for mammals (no matter at what stage) then you are what you are; nothing can be added later. What might they have been if they had managed to complete metamorphosis at the right time? Well, it is interesting to note that there are four-legged, furry mammals that also have echolocation and are crepuscular. These are the tenrecs and some shrews.

7. MAN ALWAYS WALKED UPRIGHT

Man was always intelligent. Could man have been more intelligent?

We have accepted that any change into a new species is by metamorphosis. We also realise that this must occur when creatures are very small and while still requiring further metamorphosis to reach a stage where they can produce copies of themselves rather than their previous form. Mammals produce copies of themselves and that means they no longer require metamorphic changes. As mammals of any size cannot metamorphose to a different species then clearly once a shrew, always a shrew. Therefore, from this line of reasoning we can see that man could not have descended from an ape or a shrew.

Neither can changes be made at the foetus stage by the mother's behavioural actions. We read about many strange ideas that were held, throughout history, concerning such inheritance but these have long been dismissed as nonsense. After all, we don't expect that a mother constantly using a car is going to produce children with shorter legs or a mother constantly carrying heavy shopping bags is going to produce a child with exceptionally long arms. Nevertheless, strange as it may seem, similar, extraordinary and illogical statements do seem to crop up now and again and are exposed within these pages. It must, therefore, remain that the only change any mammal can make is to gradually increase in overall body size with each new generation.

If 'once a shrew always a shrew' is true then man must always have walked upright on two legs from the moment of being human but we read of many fanciful ideas as to why early man became bipedal. These ideas range from: freeing the hands for tool making, coming down from the trees onto the savannah, finding it easier to spot predators, acquired from walking upright on tree branches, and even those book illustrations that show man gradually evolving from the stooped position of an ape to an upright bipedal one. But, as man did not evolve from an ape, followed by various stages of half-ape, then his walking never needed to evolve from a stooped position. It makes more sense that the ability to walk upright was there from the beginning because of his bone structure. Man, unlike apes, has a small pelvis with a backward bend (the sciatic notch) that allows him to walk upright without effort. If we really had started off with an ape's pelvis, then of course, we would still have an ape's pelvis. Even if the Darwinian idea, that changes are made by mutations over thousands of years, had been correct, the structural problems involved with changing an ape's pelvis into a human pelvis would have made it impossible. However, it can easily be explained by metamorphosis. We can see that metamorphosis to man must have completed at an early stage while the pelvis was still rounded and upright and before it had a chance to continue its growth further and elongate as in the ape. (This is referred to again in a later chapter.) Man's bone structure would have meant that walking upright on two legs was the easiest and most comfortable way for him to move around on the ground. It would be impossible for man with his human pelvis to walk in a half-upright position for any length of time. Try it and you'll find that your back will ache unbelievably and quickly become beyond endurance. As we have said before: *A new life form or species straight from its metamorphosis, establishes itself quickly as a functioning, workable body. It uses whatever it has been endowed with, to the best of its ability and in whatever way comes most easily and comfortably to it.* In man's case it would have been, undoubtedly, to walk upright.

Not only would the upright gait have been there from the moment of becoming the new human species but also the intelligence would have been there too. In fact, we would have been completely human from the start. If due to some terrible catastrophe we were left without the trappings of modern civilisation, with all its achievements and store of knowledge, our lives would probably revert to being much like early man. Our intelligence would not have changed but we would be back on the first rung of the ladder and using our intelligence to tackle our immediate survival in much the same way as early man in his struggle for existence. There is much discussion about very early man having a smaller brain but surely, if you go back far enough, man would have been much smaller anyway. If man was smaller he would have had a smaller skull and therefore a smaller brain; not less developed, just smaller. So, now we have raised a question; is a small brain less intelligent and a bigger brain more intelligent? This might not necessarily be the case and, of course, it depends on what you call intelligence. If we look at small creatures, that obviously have small skulls that can only accommodate small brains, they don't seem to have less intelligence than some larger creatures with larger brains. For example, a crow can hold a stick in its beak and manipulate it as a tool to extract insects from crevices, but we would certainly be astonished should we observe a cow with a stick in its mouth as a tool for scratching, or swatting flies. Man had an intelligent brain from when he emerged from the metamorphosis of his previous form. In the next chapter we will dig back even further to find out what form we took before our final metamorphosis to man.

We must not forget that the brain would always have had the ability to develop. Even today, when we learn new things, permanent changes are made in the brain. Therefore, the intelligence of early humans would have advanced rapidly as they learned to use language, make tools, make fire, and work out ways to protect themselves. Also, with the development of language they had the ability to exchange knowledge and work out problems together.

Before we leave this subject of intelligence the question begging to be asked is whether humans could have been more intelligent. The information we have gathered so far suggests that it is unlikely that man could have developed further during the metamorphosis from his previous form and become more intelligent. This is because if more time had been available for developing the brain then the extra time would have resulted in other developments such as fusing of the frontal bones and elongation of the facial region (like in the ape). Alternatively, if metamorphic development had been given less time would it have produced a human-like creature with higher intelligence? It is difficult to see how this could happen without remaining mostly in the previous form. Development would still have progressed far enough to have lungs and the four-chambered heart, but the facial region would be flatter with no protruding nose (i.e. nostrils back in the face). The external ear would be non-existent and the arms shortened with the possibility of webbed fingers and toes. Unfortunately, the extremely flattened facial region, nose and jaw, would result in drastic changes to the mouth and throat regions and this probably would have made speech impossible. Thus it is difficult to imagine how such a creature could manage to advance much beyond this state, or benefit from a larger brain, or indeed be called human.

8. THE PREVIOUS FORM OF PRIMATES

Amphibians take to the trees. The Tarsier.

We know that evolutionary biologists are in agreement that life began as microscopic life and gradually evolved into the creatures we see around us today. However, no one appears to have considered exactly what form life took between microscopic life and the first mammals. In fact, not even the previous forms of the present species, particularly primates, have been given much attention. Apart that is, from the often churned out incorrect ideas of small shrew-like creatures or half-ape, half-human creatures. Why have ideas been so thin on the ground to cover this vast expanse of time from our beginnings as microscopic life? Maybe the belief that we have developed from the shrew and the ape has reduced the need to probe any deeper. Or maybe the belief that evolution took place by small changes over millions of years have made it too mind boggling to contemplate or too distant in the past to ever know the truth. Can we now find answers? If man did not develop from a shrew-like creature or an ape, then what was man or indeed any primate before they became the new species?

We are now probing back into the dark area of that unknown, distant past of our world and our beginnings. Wherever this line of enquiry leads it will at first sound like nonsense because we need time to assimilate and accommodate completely new ideas. The idea that any primate, however small, could have begun its life by metamorphosis from a completely different creature, would probably be instantly dismissed as impossible. However, it is even more impossible to be a microbe one moment and a primate the next. There must have been intervening creatures. These would not be the so called 'missing links' searched for since Darwin's 'On the Origin of Species' in the hope of showing conclusively that new species were made by mutations over millions of years. No, these intervening creatures would certainly not be mammals but very small, completely different creatures that were still capable of change by metamorphosis. Our logical deductions have shown that it cannot be another mammal because mammals are complete on metamorphosis. It must be a creature that is very small, still needs metamorphosis to change to a state for reproduction and does not produce copies of itself but its previous form. Therefore, from this we are quite clearly told that it can only be an amphibian because there is nothing else that fits the bill. Your first reaction will probably be to exclaim, 'You must be joking,' or 'Really? And which amphibian am I supposed to be related to?' Well, the most likely candidate is the frog. In Darwin's time people were horrified to think that they had descended from an ape. So today, is it better or worse to be told that you have descended from a frog? Alas, it takes more than a kiss to turn a frog into a human prince but the old fairy tale certainly had a grain of truth.

Whereas today's tale of a shrew changing into a man by chance mutations over millions of years can be nothing but a fairy tale. Everything about a shrew would need changing, requiring endless mutations, and it goes without saying that present day shrews are not showing any signs of change. Metamorphosis is triggered in the tadpole stage and the frog is an adult form for reproduction until the frog tadpole can manage to take metamorphosis on to the primate. The frogs of today, of course, originated from a very much later arrival of life on Earth than those responsible for today's primates but they are the precursors of the next line of primates. It is interesting to note here that the frog has always been a favourite for dissection by university students because the frog's internal organs are arranged similarly to the human body and are an excellent introduction to human anatomy.

The frog tadpole can change over a period of months from a gill-breathing fishy creature into a lung breathing land creature with legs. There, at the moment, the development ceases and it breeds in this stage we call the frog but it is only a stage because in most cases it cannot produce a replica of its own kind; it has to produce its previous form – the tadpole. It is also obviously incomplete, no outer ear, incomplete pelvic girdle, incomplete skull, the heart is three chambered and not four chambered like birds and mammals, the ribs are either absent or virtually unossified, the thorax is undeveloped and the skeletal system of the fore limbs is not joined to the thoracic vertebrae. When conditions are right, and this means climate/temperature, possibly light levels and moisture levels, required nutrients and minerals, development/size of the creature involved and maybe other requirements, then complete metamorphosis can be achieved. Changes in the tadpole will then be triggered (genes turned on that were previously not used) to go on past the frog stage to the primate. This will involve further development of the pelvic girdle enabling support for an upright position on those long back legs. (Well-developed leg muscles for the task were pre-developed in the frog stage.) There will be an increase in brain size that will be added to the primitive brain stem and will increase skull size upward above the eye sockets. The increase of skull size, commencing above the eye socket, will in many cases leave a brow ridge. This increase will also bring the eyes more forward facing. The ear will continue to develop and will produce the external ear (pinna). The heart once two chambered in the tadpole and three chambered in the frog will continue development to the four chambered heart. Hair will commence growth and, of course, the change to a mammal will mean development for the womb and mammary glands. Due to changes triggered to go on past the frog stage, some parts that were necessary for the frog will be bypassed as no longer necessary (i.e. junk DNA).

There are many changes to be made but by metamorphosis these changes can be made easily and smoothly. Many were partially built up in the system in readiness during development to the frog stage. Also think how many changes were necessary to change the tadpole into a frog. Now think how impossible it would be to achieve all this by small, random changes made by mutations over millions of years. Also, such random changes would be without any prior development in place to build onto and this makes it an impossible task. A hypothesis that shows changes are made by DNA during a process we call metamorphosis is surely more scientific than one based on chance mutations and definitely more logical. Also, of course, this method can make large changes in a reasonable time scale. One more point is that the length of the DNA in a frog is more than twice the length of the DNA in man. Isn't this because frog DNA contains all the information ready for those changes by metamorphosis to primates?

There is one primate that can clearly show its frog ancestor connections. This is the Tarsier. The Tarsiers are primitive primates so you would expect them to exhibit more of their frog ancestry. They have similar pelvic bones to the frog and leap in the same way on the ground with powerful hind legs. On their fingers they have fleshy pads like the Tree Frogs which help them to grip in the trees. They also blink their eyes when swallowing. (In frogs the blinking of the eyes pushes the eyeballs down and helps them to swallow food.) Both frogs and tarsiers have fused tibia and fibula.

The rainforests of the world are home to many tree frogs. It would seem very strange that amphibians have taken to the trees but for the knowledge that in their next (and last) metamorphosis they will become primates. Most primates, of course, are naturally at home in the trees. There are many different types of frogs, including toads, in many different parts of the world. It is during their tadpole stage that these will eventually make further changes, through metamorphosis, to many different types of primates.

Fossils show that the skeletal shape and body plan of the frog has remained almost unchanged over the last 190 million years. This is what we would expect – all frogs from ancient fossils through to the present day won't show any signs of change because the only change possible is by metamorphosis into a completely different creature. We will never find small changes made by chance mutations, not even over millions of years. It just doesn't happen. Also, today's frog is not a direct descendent from those frogs of 190 million years ago. Today's frog developed very much later in time, from another new cycle of microscopic life that drifted down onto our planet. These continual drifts of microscopic life eventually produce the same or very similar forms and will undergo the same metamorphic transformations.

9. WE DIDN'T EVOLVE FROM THE APES

Why man is so different. Neoteny.

So, if primates descended from amphibians by metamorphosis, how is it that man is so different? To answer this question we have to look at metamorphosis again. Climatic conditions could either prevent metamorphosis altogether or could speed up or slow down or halt the process at any stage of development. As we said before, metamorphosis is a risky business. We also established that this is how we have sea mammals. They were once very small creatures that were in the process of change but, due to climatic conditions, their metamorphosis ceased and they were trapped to a life in the sea because they had not developed far enough for life on land. It is logical to assume, therefore, that the metamorphosis to man must have been brought to an early completion point. We only have to look at ourselves to see why. We have retained the smooth, supple skin with fine or little hair. Our pelvis has stopped growth at the rounded, upright position that enables us to walk upright. Our facial region also shows that growth was arrested early as it has remained flat and not elongated as in the ape. The flattened face meant a poorly developed sense of smell but this was a small loss compared to what we gained. Imagine if the completion time of our metamorphic change had been extended for any length of time and given more time for development; what then? This could only have resulted in the continued, outward growth of the nose and jaw to the detriment of the skull which would have been pulled and elongated forward. The arms would have continued growth and more or thicker hair would have been produced and the skin dryer or more wrinkled. The pelvis too would have continued growth, causing elongation and eliminating any hope of walking with an upright two-legged gait as a main means of locomotion. So, yes, we would have looked like an ape. It can now be seen that we didn't evolve from the apes; the apes are the continued growth of the human frame.

If we think about growth to the facial region we can envisage that the human nose would have been the beginning of the growth of the forward projecting face and jaw of the ape. The human nose is mainly cartilage showing that it was an area with potential for growth. The growth and elongation of the nose and jaw area pulls the skull forward, flattening and reducing cranial size. The lengthening of the jaw to the detriment of the skull is seen in other animals too. We must assume that in these animals, as the jaw elongated forward, it brought the teeth buds with it. Depending on how much the jaw extended, it is likely that some teeth buds would be carried forward while others are left behind. In some cases this might create a long gap between teeth at the back and teeth at the front. As the jaw was not extended in man, the teeth were unable to move forward and this would account for why the wisdom teeth are so far back.

The human infant has a high forehead, small nose and flattened face like an adult human. Its skull is made up of bones that are not completely joined together at birth. This enables the brain to expand, with a rapidly increasing brain size after birth. Somewhere between the ages of one and two years the two parts of the frontal bone fuse into one and the metopic suture (a line that marks where the two sections of the frontal bone come together) disappears. In the ape this fusing of the frontal bones occurs while still in the womb, so although the ape infant starts off with a high forehead, small nose, and flattened face at birth, its frontal bones are already fused so the brain cannot expand in size. Growth then pulls forward the facial region and creates the low forehead and protruding jaw.

If we go back to the time of metamorphosis, from the amphibian frog stage, we can see that metamorphosis ending early for humans meant there was insufficient time for the frontal bones of the skull to fuse. This was fortunate for us because it enabled the brain to enlarge after the end of metamorphosis. With the ape, the closure of the frontal bones happened during its lengthened metamorphosis – so there was no hope of any further brain growth after metamorphosis ended. Therefore, following on from this, we can see that the timing of the fusion of the frontal bones is then included in the DNA and re-enacted in the development of the foetus in the new species.

Those that believe man descended from the ape (a fact we now know is wrong) use the term 'neoteny' with regards to man. This basically means that somehow the retention of the infant ape stage was achieved and resulted in the human species (i.e. giving us the high forehead, small nose and flattened face). In a sense this is true because the metamorphosis to man ceased at this time whereas it continued on past this stage to produce the first adult ape. So it was all down to the amphibian metamorphosis, where for us it ceased early and in so doing prevented further development and produced the human being. When metamorphosis did not cease at this stage, but was able to continue, the extra time for development resulted in the ape.

The summing-up of this chapter leaves us in no doubt that the metamorphosis from the amphibian (frog) stage was programmed to produce man. However, climatic conditions on this planet, (such as global warmings and ice ages) were capable of disrupting and prolonging the process. The completion and shutdown of metamorphosis at the intended time (early stage) would have produced man, whereas the unfortunate lengthening of the process of metamorphosis would have continued man's growth, turning him into the ape.

The lengthened metamorphosis gave time for the frontal bones of the skull to fuse together, preventing any further expansion and development of the brain. The extra time also pulled the jaw and skull forward. It elongated the pelvis – preventing an upright gait as a main means of locomotion. The arms became lengthened and there were changes made to the hands and feet, plus thicker body hair etc. So, instead of producing man, the final result was the ape – and once you are what you are, there can be no further change and no going back. We need to embrace these creatures as part of our family, because they are more human than we ever realised.

If you are searching for man's origin by looking back in time for a creature that is like an intelligent ape, then you are going to be disappointed. We now know that the origin of man never arose from small changes to an ape over long periods of time – an ape will always be an ape. You need to look for a completely different creature – one that can make a metamorphic change to produce the first man. The first man will arise from an amphibian. He will be small, about 50cm tall, and certainly walking upright on two feet. His head will obviously be small (in keeping with the size of his body) but will contain an intelligent brain. For example, Homo floresiensis might be listed down with the chimps for average brain size but he would, nevertheless, have been more intelligent because his jaw and skull were not drawn forward. This would indicate that fusing of the frontal bones didn't happen until sometime after the end of metamorphosis – he was human.

A rise of temperature on the planet would have triggered metamorphic change in the amphibians and given rise to many different types of humans in various locations around the world. Some of these tribes would doubtless have come across each other but due to the aggressive streak and territorial boundaries in male mammals, it is highly probable they would have attacked and killed each other. It is also probable that the women were taken by the victors because this would explain the Neanderthal and other genes within us.

10. THE SKELETON

The insect stage. Do we have one common ancestor?

By the time the very earliest life forms had evolved through to the insect stage there appeared the formation of the skeleton. This formed on the outside of their bodies. (The exoskeleton made of chitin at this stage.) This was obviously very useful because it prevented the drying out of small vulnerable parts and allowed freedom of movement away from water or damp locations. It was also laying down the foundations for the skeletal structure that would eventually be needed for metamorphosis to higher forms of life. Once again we can see how the beginning of a development for a further stage is not a hindrance in any way but of positive use. Insects are not a completely separate form of life that just happened to evolve their skeletons on the outside. Insects, as well as crustaceans, are all a previous stage of more complex forms. Therefore, they have parts developed in readiness for future metamorphosis. Life in the insect stage lives inside its skeleton and this was realised long ago by Etienne Geoffroy Saint-Hilaire, the French early nineteenth century leader of transcendental biology. He correctly identified the external skeleton of insects with the internal bones of vertebrates. He interpreted each insect segment as a disc of the vertebrate spine and argued that insects dwelled within their own vertebrae. This was a great step forward but unfortunately, it was never recognised or accepted during his life time or indeed since.

We not only see the development of the eye from primitive life forms to insects and eventually to higher forms of life, but also the way in which the skeleton has been gradually developed in stages too. During the process of metamorphosis from the insect stage, to the new form, we could imagine creatures being turned almost inside out because all the soft parts of the body, that needed to be redeveloped, needed to be transferred to the outside of the skeleton. However, we have to remember that metamorphosis is triggered from the larva or caterpillar stage. These are mostly soft bodied so to produce the early beginnings of the skeleton on the inside, from the inside, greatly simplifies the matter. The brain, however simple, always remains intact during the process of metamorphosis although it is developed further to be appropriate for the new form. Connections with our humble past can be seen in our own body plan. For example, our ventral nerve cord is protected by our spine and our brain is encased in our skull. We can see from the human brain that we have a basic stem onto which further development was added later.

The same basic skeletal plans, body parts and organs are used in most creatures. Does this mean that we all come from one common ancestor? Or does it mean that the microscopic life that descends to Earth contains the same body plans within its DNA, although with some tweaks here and there, for its forthcoming construction of life? Certainly when we think of the final metamorphosis into creatures such as mammals, birds and reptiles we are taken back to the amphibians as their previous form. This is because we know that amphibians are the last in line of the very small, immature creatures that are still capable of metamorphosis to a state for reproduction or to a new species. So, progressing from this, we can say that mammals, birds and reptiles do have one common ancestor – the amphibians. But can we go back further than this? If the microscopic life (bacteria) is always carrying the same DNA to develop life from simple to complex forms, as our cycle of life suggests, then it means that the same genes will have control of the development of the same body parts in all forms of life. Nevertheless, they can only switch on the appropriate, further stage, of each body part, as creatures progress through the simple to complex forms by metamorphosis. So, you could say that life's one common ancestor is right back at the beginning – the bacteria.

Microscopic life descending to Earth would not only fall into lakes, ponds and other waterways but also into the vast oceans. This would present difficulties much later when they reached the potential of the insect stage. Crustaceans appear to be insects in the sea or maybe we should call insects crustaceans. Like the land insects they also have their skeletons on the outside. However, as insects in the ocean they would never be able to metamorphose into flying insect forms even though they could go through other various stages of change. The flying stage would be impossible for those in the oceans because without plant stalks reaching up above the water the larvae would have nothing to climb up in order to metamorphose into a flying stage. This is essential if you are changing to breathing air and have wings that need to be unfurled and spread out to dry.

When we look at ancient insects found in amber we see that they are practically the same as the insects around us today. Beautifully preserved in their amber tomb, from the Jurassic and Cretaceous period we find bees, mosquitoes, flies, ants etc. Even after all these millions and millions of years, the insects of today are so similar to those in amber, that it must prove, without doubt, that each new cycle of life from space comes with the same instructions. This is the only explanation for insects of today being identical to the insects in amber because they cannot be direct descendants after such a long period of time. We do need to add that it cannot possibly support the Darwinian claim that life advances by small chance mutations over millions of years, not only because there are no signs of mutations but also because this system of chance could not produce identical forms.

11. THE FINAL METAMORPHOSIS

From the amphibians. The dinosaurs.

From the simple life forms to the insects and finally to the amphibians, changes are made by metamorphosis. After that there are no more changes by metamorphosis because the resulting creatures are capable of breeding and replicating their own kind and no longer require a stage for reproduction. The metamorphosis of the amphibian larval stage produced the mammals, reptiles and birds but these resulting creatures cannot metamorphose. In fact, the only change these creatures will make is a gradual increase in size. Although reptiles and birds are able to produce their own kind, instead of producing their previous amphibian form, they are in many ways incomplete in their development. Birds lay eggs, have a cloaca, and no outer ear but nevertheless, possess a four-chambered heart and are warm-blooded. Reptiles, with a few exceptions lay eggs, have a cloaca, no outer ear, cold-blooded and only a three-chambered heart (except for the crocodile, although its four-chambered heart is a little different). However, incompleteness can also be found in some mammals. We find this in the sea-mammals, the marsupials and the monotremes. Metamorphosis to primates, from the frog line of amphibian larvae, has been discussed previously.

When the salamander (that includes the newt) line of amphibian larvae metamorphosed into lizards, there were times when climatic conditions prevented a smooth and complete change. For some, their metamorphosis ended too early and they never reached the stage for the development of legs. These were lizards without legs; the snakes.   For those where conditions were ideal for metamorphosis, a smooth, complete change produced the lizards.   However, for those where climatic conditions slowed down the process (rather than ceasing it) and so delayed the completion of metamorphosis over an extended length of time, we have the birds. Birds are therefore, the continued development of the lizard. This extra time, for metamorphosis to continue, allowed the development to a four-chambered heart, growth of scales to feathers, elongation of hand and finger bones, growth of a horny beak over the jaws and continued growth to produce large collar bones. As the hand and finger bones continued to grow, (instead of completing at the usual forelimb length with its digits,) the adjoining skin remained because it missed the time slot for retraction from between the fingers.

The evidence, for early cessation of metamorphosis being responsible for producing snakes instead of lizards, is as follows: Lizards have two lungs but a snake has only one lung; the other one is either non-existent or non-working. Snakes do not have eyelids like lizards but a transparent unblinking covering. Some snakes have rudimentary leg bones or pelvises in the abdominal cavity. Snakes do not have hearing apparatus. Therefore, it can be seen that had metamorphosis continued, the lungs, eyelids, legs, and hearing would have continued their development and produced the lizard.

Way back in prehistoric times when tiny lizards had metamorphosed from one line of tiny amphibians they began, generation after generation, to gradually increase in size until eventually they became the huge monsters we call dinosaurs. We tend to club together all these large creatures as 'dinosaurs' but it is obvious now that they were not all huge lizards; some were huge birds. We have partly acknowledged this fact already by dividing Dinosaurs into two orders; the Ornithischians (bird hipped) and Saurischians (lizard hipped.) Unfortunately, it is not this clear cut and there appears to be a great deal of argument and confusion over this classification. A recent discovery that some dinosaurs had feathers and were warm-blooded would appear to be evidence for the huge birds. The lack of feathers and cold-blooded would be the huge lizards. How can we tell if these creatures had warm blood or not? Well, we need to know if they had a four-chambered heart. A four-chambered heart is characteristic of warm-blooded animals. In 1993 a unique fossil dinosaur (Thescelosaurus) was discovered with the heart still preserved. This heart had four chambers and only one aorta. If this information is correct, then it was a warm-blooded bird (birds have a four-chambered heart while lizards have a three-chambered heart). This fossil was also bird-hipped.

Our 'cycle of life' clearly shows that our present large birds, such as the Ostrich, Rhea, Cassowary and Emu, and the large lizards, the monitor lizards and the crocodiles, although not huge at present, will one day end up like the dinosaurs. Also, there are not so many of them in our present cycle because many disasters on this planet have been responsible for their extinction long before they could reach this size. Likewise, the tiny new generations of lizards around us today are ready to begin another new cycle of life but are rapidly being wiped out through loss of habitat and pollution.

We now know that mammals, reptiles and birds never go through metamorphosis and therefore can never change into anything else. Therefore, we know that the previous hypothesis proposing that animals gradually changed by slight mutations over millions of years, into other animals, is completely wrong. This also means that the current idea, suggesting that not all dinosaurs became extinct because some of them changed into birds, is also wrong. However, they were close to the truth and might have taken it further given the knowledge that metamorphosis is the solution to evolution. The connection they are making, with birds from the dinosaurs, is based on the evidence they observe but their timing of the event as well as the process is wrong. To explain this connection we need to go back to the time when dinosaurs were very small; to their beginnings as tiny lizards. As we saw earlier in this chapter, when metamorphosis did not complete at the lizard stage but continued to develop onwards from the lizard form, it produced the bird. So, this is when the lizard (dinosaur) to bird connection happened, at this very early stage, and not when the dinosaurs (lizards) were huge. Obviously the tiny lizards and birds then continued to grow bigger and bigger over time until they eventually became the huge lizards and birds that we now group together under the name of dinosaurs.

Way back in 1868 Thomas Henry Huxley argued that, 'birds and reptiles were descended from a common ancestor'; we can now agree with him. However, because we have introduced the process of metamorphosis we can show that the lizards (dinosaurs) and birds not only had a common ancestor but were very closely connected together in their development. Thus we have the problem of working out which dinosaurs are more lizard-like or more bird-like. The length of metamorphosis not only produces differences but can also mean there is some overlap of characteristics that prevents a neat splitting into separate groups. (And we do like to put everything into neat compartments.)

12. STRETCHING THE IMAGINATION

Long beaked birds and long trumpeted flowers. Is mimicry really mimicry?

Some far fetched ideas have arisen as a result of the belief that changes are made by mutations over millions of years. We need to look at these and discuss them logically. We are told that if there is a flower with a long, thin trumpet that we will find a bird with a beak long enough to reach the food supply and to pollinate the flower. So far so good, but then the argument goes that this is a close relationship that has evolved over millions of years for mutual benefit. (i.e. that trumpets or beaks grew longer.) It is not difficult to see that where there is a long trumpeted flower that a long beaked bird will take advantage of this food source and thereby pollinate the flower. If insects do not appear to be responsible for pollination then it's a safe bet to say that it is pollinated by a long beaked bird. It is obvious that with the variety of insects and birds there will always be something that takes advantage of a food source neglected by others. However, no flower could have hung around for millions of years waiting for a bird to slowly evolve a beak long enough for the job. As for the flower, why would it need to evolve a long trumpet for long beaked birds when, without the trumpet, it would have plenty of insects and birds quite capable of providing this service. The lack of logical reasoning here falls into the realm of the ridiculous. Beaks do not slowly evolve to reach food sources in flowers or to catch insects in deep crevices. Birds possessing long beaks use them to their advantage. If they didn't have long beaks and couldn't access a food source they wouldn't hang around but fly off and seek food elsewhere. There is similar strange reasoning attached to the extinct Huia bird of New Zealand. It is said that this bird evolved a long beak for the purpose of catching insects in crevices. However, only the female had a long beak which proves this thinking is wrong. If the long beak had evolved for the specific purpose of catching insects in crevices it had to be in their genes to pass on and that would mean the male bird would also have had a long beak. As this condition affects only one sex the cause for the difference must be attributed to the sex chromosomes. The idea that body parts can be changed to meet demand keeps sneaking in by the back door but it is an old theory that has been toppled many times. Anyway, by the time any part could manage to change, by chance mutations over many millions of years, we can safely say that the demand for change would no longer be required. (i.e. the long trumpeted flowers would have died out or the birds craving for long beaks would have died from starvation or simply found something else to eat.)

Let us consider some ideas often quoted as mimicry. No matter how much we might desire to look different and no matter how much it might benefit us to do so, it is just not within our power. Hoverflies are quoted as achieving a striking mimic of the wasp in order to avoid being eaten by birds. But how could hoverflies possibly know that birds were not eating wasps or know that birds were avoiding wasps as unpalatable by their different colours or patterns? We are expected to believe that the hoverflies changed over a vast period of time and that only those more nearly resembling the colouring of wasps survived to breed. Does this really make sense? If it was possible to change, and assuming that all insects were spending time watching wasps, wouldn't all insects have ended up the same colour to avoid being eaten? Also, if they had managed to survive over vast periods of time without looking like a wasp then surely the change wasn't necessary. Can we explain these similarities in any other way? As creatures cannot change by desire then we must seek logical reasons for these similarities. We could assume that Hoverflies had the same colouring as wasps in the first place. This would take us back to their emergence from their last metamorphosis. There is a possibility that a problem arose during their metamorphosis. This could have resulted in variation of form but still maintaining colour and pattern to some degree. There is some evidence for this as apparently some Hoverflies have been known to push the tip of the abdomen into your skin, when captured, even though they cannot sting. This suggests that part of the mechanism for stinging is in place but not fully developed. If their metamorphosis had continued, they would have completed their stinging capability and become wasps. It is also possible that a food source is responsible for the colouration and poisons. Caterpillars that feed on certain plants not only contain these substances but retain them in their butterfly stage. Ragwort produces toxins but they do not affect the caterpillar of the Cinnabar moth that feeds on it. The caterpillar stores the poison (which maybe is how it avoids being poisoned by the Ragwort) and this is still retained during the transformation to the adult moth, making it distasteful to predators. So, creatures that are credited with being mimics may have a simpler reason for their patterns and colouration.

An often quoted idea is that colour is there to warn predators that they are poisonous to eat. Nature doesn't usually go around ringing warning bells for the benefit of other creatures. This is completely illogical or, at best, fanciful thinking. It is probably chemicals within their bodies that are responsible for their colour as well as making them poisonous and distasteful. We are told that the Fire Bellied Toad exposes its red or orange speckled underside towards its enemy as a warning. However, it is far more likely that it is trying to protect itself from attack by exposing its venomous skin secretions towards its enemy. Another example is the often quoted Tiger Swallowtail butterfly, Papilio Glaucus, a presumed mimic of the poisonous Pipevine Swallowtail butterfly, Battus Philenor. Apparently the female can have two forms; the usual yellow coloured one that is similar to the male and one that is brown to black in colour. The dark one apparently resembles the poisonous Pipevine Swallowtail. This really is stretching the imagination. We are even told that some poisonous butterflies mimic to resemble other poisonous butterflies to make extra sure they are not eaten. This is utter nonsense. One couldn't hope for a better example to topple the idea of mimicry. It is interesting here to refer back to our thoughts on the Peppered moths. Some of these moths were supposed to have changed to a dark form and were camouflaged on sooty tree trunks, thus avoiding predation. We found that temperature was a highly likely reason for this event. Likewise we find that the female Tiger Swallowtail butterflies that turned black were mainly in the south. The south would be warmer so again it looks like temperature is the cause and nothing to do with mimicry. Sometimes having the same colour can be due to ingesting the same chemicals from plant food, or it can be due to temperature triggering changes, or even differences during metamorphosis. However, it can never mean that some creatures have set about changing their appearance because of their desire to mimic another creature. In many ways mimicry is that same old idea creeping in again where pure desire can cause beneficial changes (although how much a Hoverfly can desire is surely questionable).

13. LIFE IS NO ACCIDENT

We come complete with instructions. A speck of dust capable of asking questions.

We often read that the thickened skin on our feet presents a puzzle. It is thought miraculous that the body appears to anticipate the need for the thickened skin before it is required; even before a child is old enough to walk on its feet. Yet the feet are there for walking on, so surely the thickened skin would be a built in necessity that comes with the development of the feet. You could just as easily say the same for muscles in our legs before they are needed or eyelashes to protect our eyes from future dust or immune cells to protect us from future invaders. The situation would be hopeless if we had to wait for small mutations over millions of years. The body has to be equipped in advance because it cannot suddenly activate changes when the need arises. The real puzzle is that we are such a complete package and don't require any changes by mutations to improve our bodies. As we are inside this body it's possible for us to think of one or two things that would have been an improvement from our point of view. However, if we were able to reassemble our parts on the drawing board, we might well find that there was no other option to what we have, without losing or jeopardising something else. This is because development through metamorphosis not only had to produce bodies that worked efficiently but also bodies and organs that could easily accommodate the alterations that would be needed for further metamorphosis in the future. The whole sequence had to be constructed in such a way that it would be possible to easily add, extend or modify parts as programmed in their DNA. For example, the heart is two chambered in the tadpole and during metamorphosis to the frog it develops into a three chambered heart. Finally, when metamorphosis is able to continue on to the primate it further develops into the four chambered heart.

Was life a fluke? We have worked out that life was not a chance occurrence of chemicals coming together in a little warm pond. We also worked out that life originated in space and descended down to Earth. However, we still have no idea how life originated in space or indeed where it originated. We only know that it is still being churned out because it is still descending to Earth. We know that the microscopic life forms that descended to Earth were extremely lucky compared to those that descended to other planets in our solar system. Compared to them the Earth was a 'Goldilocks' planet - everything was just right. Nevertheless, nature (life) had to set up its own paradise rather than being embraced by this planet of turmoil and unrest. The one commodity it has always provided is large quantities of water and this is essential for life. Apart from the provision of water, it seems we were on our own. So, was it a mindless event without any prior planning, design or thought? It doesn't appear to be so, because we worked out that life comes with all the instructions in its DNA for development through metamorphosis from simple to complex forms. In fact, we appear to come fully equipped for every eventuality. However, it would appear that life isn't directed carefully to habitable planets; it's a very hit and miss affair with more potential life being lost than those that find a habitable planet. Interestingly this is comparable to seed dispersed by the wind. Also, bacteria and viruses etc. are still descending to Earth even though they cause many diseases in the advanced, complex life that is now well-established here. But without this, there could be no cycle of life and therefore no repopulation of a planet once its creatures had grown large and become extinct. There are so many questions and we are desperate for answers. If only we understood the universe. If only we knew what life really is, where it comes from and why it is still being churned out in large quantities; then perhaps we might understand what is going on and how we came complete with instructions (DNA) to build and equip us for life on a planet like Earth. Something somewhere must be mass-producing all the life (bacteria) that is constantly drifting around in space and no matter what it turns out to be, it can only mean one thing; life is no accident. Therefore, are we an integral part of the universe?

When we try to answer questions like these, we tend to look to the skies and into space for the answers. We look to something that is bigger and greater than us. Looking into space is awe inspiring. We realise just how very small we are when we compare ourselves to the scale of what is out there. Although we can now see into space further than ever before, back in time to some of the earliest galaxies, we have not found answers to the existence of life. It all seems huge and hostile to life; like falling into the strange workings of some colossal machine. If we look the other way, to things that are much, much smaller than us, (simple organisms, bacteria, etc) we are surprised to find that there are strange systems going on right under our noses. We originated from the very small microscopic life so maybe this is where we should be looking for the answers to these difficult, and at the moment, unanswerable questions.

Man appears to be a minute speck of dust in a vast universe yet nevertheless is a highly complex system with a brain capable of looking at itself and asking searching questions. It is desperate to know where it is, what it is, why it is, how it is, and from whence it came. It continues to call out across the universe but there is no reply.

Copyright © 2014 by W.J. Corkett

Revised Edition 2019.
All rights reserved.
ISBN: 9781092819732

www.ingramcontent.com/pod-product-compliance
Lightning Source LLC
Chambersburg PA
CBHW072230170526
45158CB00002BA/826